Table of Contents

We Need Light

Open the curtains! Turn on the lights! Light lets us see our world. Without light, it is hard to find our way around.

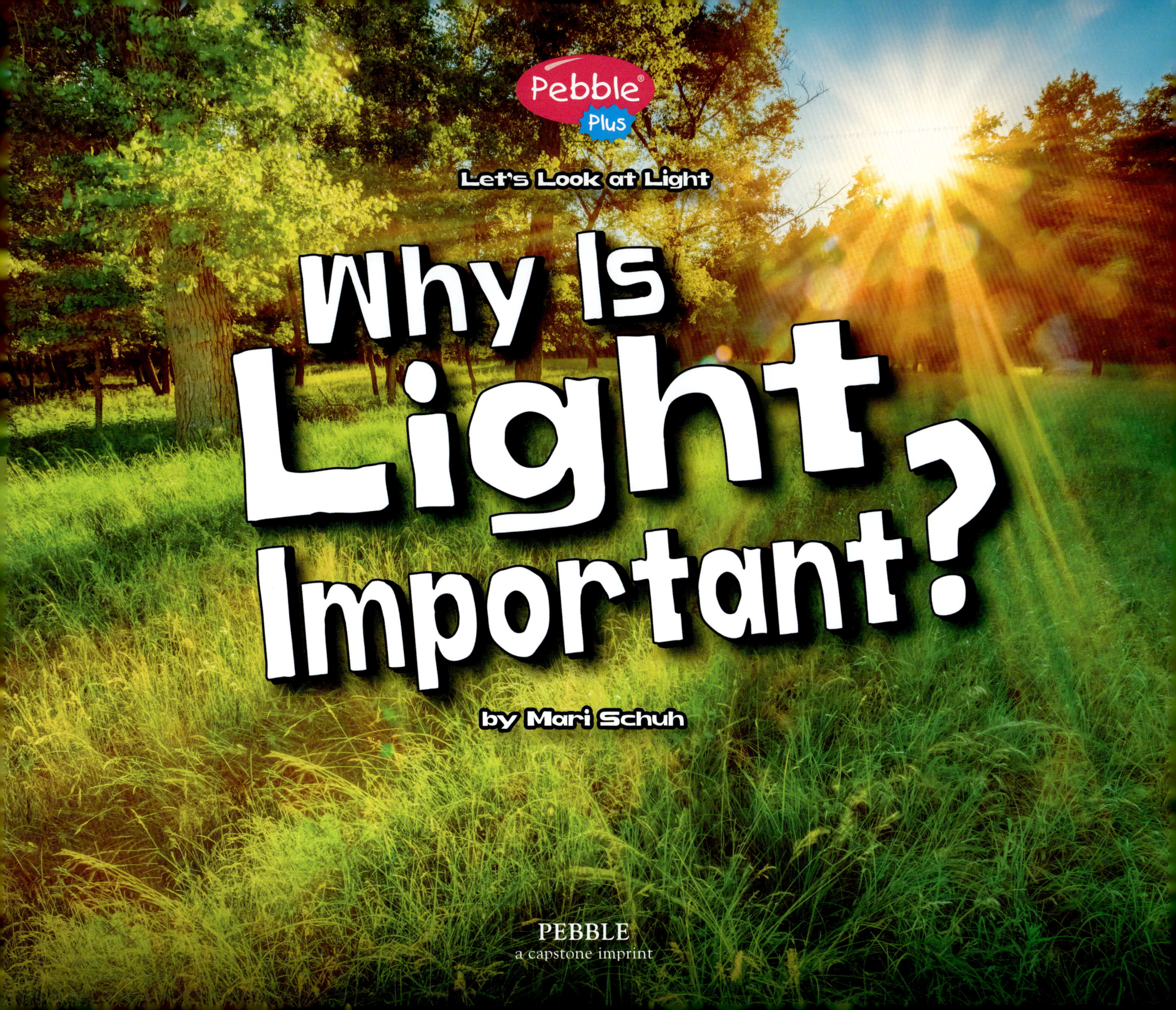

Pebble Plus

Let's Look at Light

Why Is Light Important?

by Mari Schuh

PEBBLE
a capstone imprint

Pebble Plus is published by Pebble
1710 Roe Crest Drive, North Mankato, Minnesota 56003
www.mycapstone.com

Library of Congress Cataloging-in-Publication Data
Names: Schuh, Mari C., 1975- author.
Title: Why is light important? / by Mari Schuh.
Description: North Mankato, Minnesota : Pebble, a Capstone imprint, [2020] | Series: Pebble plus. Let's look at light | Audience: Ages 4-8. | Audience: K to grade 3. | Includes bibliographical references and index.
Identifiers: LCCN 2018060252 | ISBN 9781977108968 (hardcover) | ISBN 9781977110398 (pbk.) | ISBN 9781977109002 (ebook pdf)
Subjects: LCSH: Light--Juvenile literature.
Classification: LCC QC360 .S3837 2020 | DDC 535--dc23
LC record available at https://lccn.loc.gov/2018060252

Editorial Credits
Karen Aleo, editor; Kyle Grenz, designer; Tracy Cummins, media researcher; Laura Manthe, production specialist

Photo Credits
iStockphoto: AleksandarNakic, 19, ChristopherBernard, 11, PacoRomero, 8, percds, 17; Shutterstock: DBSOCAL, 13, Grisha Bruev, 1, Jorge Salcedo, Cover Top, Juergen Faelchle, Cover Back, leungchopan, 9, Mark Schwettmann, 15, naramit, Cover Bottom, Smileus, 7, Toa55, 21, VH-studio, 5

Note to Parents and Teachers

The Let's Look at Light set supports national standards related to light and energy. This book describes and illustrates why light is important. The images support early readers in understanding the text. The repetition of words and phrases helps early readers learn new words. This book also introduces early readers to subject-specific vocabulary words, which are defined in the Glossary section. Early readers may need assistance to read some words and to use the Table of Contents, Glossary, Read More, Internet Sites, Critical Thinking, and Index sections of the book.

All internet sites appearing in back matter were available and accurate when this book was sent to press.

Printed in China.
1654

Light from the Sun

The sun is important for all life on Earth. Plants need sunlight to live. They use energy from sunlight to make their own food.

People and animals eat plants. Plants give them energy to live. So sunlight is important to people and animals too.

How We Use Light

Some lights are made by people. These lights help us see at night. Kids may use flashlights to read before bedtime.

Lights let truckers safely travel long distances. Headlights help them see dark roads at night.

Tall poles hold powerful lights at baseball fields. It's easy to see night games. The teams are ready to play!

AutoTrader.com
NBC BAY AREA
See's CANDIES
WE LOVE BEING YOUR HOMETOWN AIRLINE

Traffic lights control traffic and keep drivers safe. Drivers slow down for yellow lights. Bright red lights tell drivers to stop.

Lights help students learn. Long bulbs light up classrooms. Bright computer screens help students see their work.

Celebrating with Light

People often celebrate with lights. Lights twinkle on Christmas trees. Lanterns glow during Chinese New Year. Lights are all around!

Glossary

celebrate—to do something fun on a special day

control—to have power over something

energy—the ability to do work, such as moving things or giving heat or light

flashlight—a small light that people carry that uses a battery

headlight—a bright light at the front of a vehicle that allows the driver to see ahead in the dark

lantern—a light or candle with a frame to protect it

sunlight—light that comes from the sun

traffic—the movement of cars, trucks, and buses on roads

traffic lights—red, yellow, and green lights that tell people and cars when to go and when to stop

Read More

Crane, Cody. *Light.* Rookie Read-About Science. New York: Children's Press, 2019.

James, Emily. *The Simple Science of Light.* Simply Science. North Mankato, Minn.: Capstone, 2018.

Lindeen, Mary. *Light.* A Beginning to Read Book: Physical Science. Chicago: Norwood House Press, 2018.

Internet Sites

Easy Science for Kids: All About the Sun
https://easyscienceforkids.com/all-about-the-sun/

BBC: Light and Dark
https://www.bbc.com/bitesize/topics/zbssgk7

Critical Thinking Questions

1. Describe one way that people use lights to celebrate.
2. Why do plants need sunlight?
3. What would happen to plants, animals, and people if there was no sunlight?

Index